A World of Shapes

Written by Dawn Brunner

The world is full of shapes.
Some have special names.

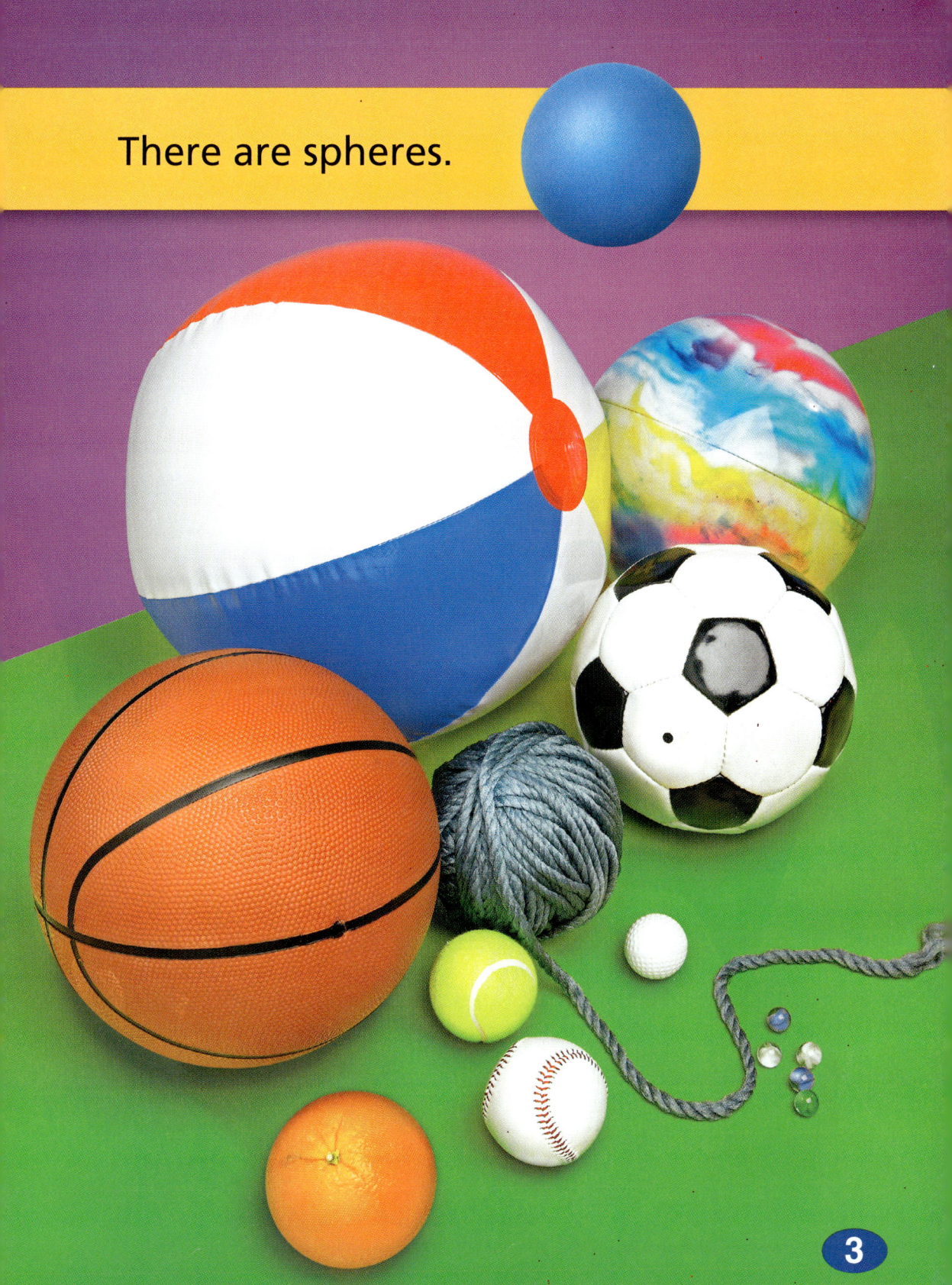
There are spheres.

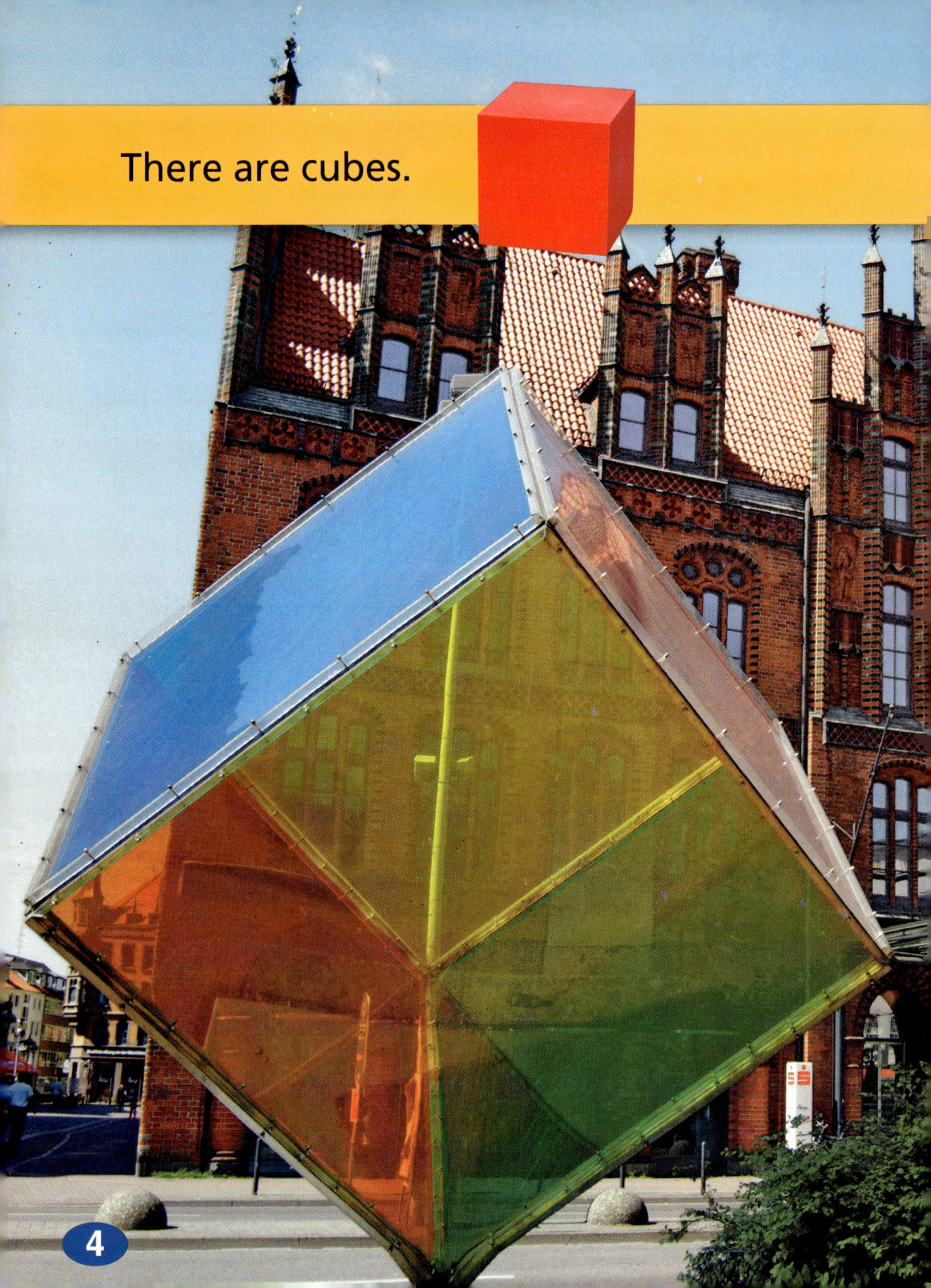
There are cubes.

There are rectangular prisms.

There are cylinders.

There are cones.

There are pyramids.

Your Turn

A playground has lots of shapes. What shapes do you see?

Some shapes are out of this world, too!